WORLD'S WORST DISASTERS

EARTHQUAKES

THE WORST IN HISTORY

BY JENNA VALE

Gareth Stevens PUBLISHING

Please visit our website, www.garethstevens.com. For a free color catalog of all our high-quality books, call toll free 1-800-542-2595 or fax 1-877-542-2596.

Library of Congress Cataloging-in-Publication Data

Names: Vale, Jenna, author.
Title: Earthquakes : the worst in history / Jenna Vale.
Description: New York : Gareth Stevens Publishing, [2025] | Series: World's worst disasters | Includes bibliographical references and index.
Identifiers: LCCN 2024000217 | ISBN 9781482466614 (library binding) | ISBN 9781482466607 (paperback) | ISBN 9781482466621 (ebook)
Subjects: LCSH: Earthquakes–Juvenile literature. | Earthquakes–History–Juvenile literature. | Seismology–Juvenile literature.
Classification: LCC QE534.3 .V35 2025 | DDC 551.2209–dc23/eng/20240201
LC record available at https://lccn.loc.gov/2024000217

First Edition

Published in 2025 by
Gareth Stevens Publishing
2544 Clinton St
Buffalo, NY 14224

Portions of this work were originally authored by Janey Levy and published as *World's Worst Earthquakes*. All new material in this edition is authored by Jenna Vale.

Designer: Claire Zimmermann
Editor: Megan Kellerman

Photo credits: Cover, p. 1 (main photo) Mohammad Bash/Shutterstock.com; series art (newsprint background) Here/Shutterstock.com; series art (fact box background) levan828/Shutterstock.com; series art (cover & caption box paper texture) Suti Stock Photo/Shutterstock.com; series art (hazard symbol) Maksym Drozd/Shutterstock.com; p. 5 Thomas Dutour/Shutterstock.com; p. 7 Dimitrios Karamitros/iStock; p. 9 VectorMine/Shutterstock.com; p. 11 courtesy of the National Archives Catalog; p. 15 Keystone Press/Alamy Stock Photo; p. 17 Frans Delian/Shutterstock.com; p. 19 EcoSpace/Shutterstock.com; p. 21 Earthquake damage in Port-au-Prince 2010-01-15.jpg/Wikimedia Commons; p. 23 중앙119구조본부 인명구조견 백두 05.jpg/Wikimedia Commons; p. 25 Doga Ayberk Demir/Shutterstock.com; p. 27 David Pereiras/Shutterstock.com; p. 29 Dino Fracchia/Alamy Stock Photo.

Printed in the United States of America

CPSIA compliance information: Batch #CS25GS: For further information contact Gareth Stevens at 1-800-542-2595.

CONTENTS

WORDS IN THE GLOSSARY APPEAR IN **BOLD** THE FIRST TIME THEY ARE USED IN THE TEXT.

SHAKES AND QUAKES

When the ground beneath your feet moves or shakes, you might be feeling an earthquake. Most earthquakes are small—thousands of them occur all over the world every year. Depending on where you live, you may not even notice them.

Other earthquakes can be so big they are considered natural **disasters**. These quakes can be deadly, destroying roads, homes, or even entire cities. It can be hard to prepare for earthquakes because scientists cannot tell when an earthquake is going to happen. There have been many large earthquakes throughout history, with the worst killing many people and causing tremendous **damage**.

DISASTER RELIEF TEAMS WORKED TO CLEAR DEBRIS FROM RUINED BUILDINGS AFTER A MAJOR EARTHQUAKE STRUCK NEPAL IN APRIL 2015.

STUDYING EARTHQUAKES

SCIENTISTS WHO STUDY EARTHQUAKES ARE CALLED SEISMOLOGISTS. THEY USE A TOOL CALLED A SEISMOGRAPH TO LOCATE, RECORD, AND STUDY EARTHQUAKES. SEISMOLOGISTS HAVE BEEN TRYING TO FIGURE OUT HOW TO **PREDICT** EARTHQUAKES, BUT SO FAR, THEY HAVEN'T BEEN ABLE TO. THEY CAN TELL FUTURE EARTHQUAKES MAY OCCUR IN A CERTAIN AREA, BUT NOT WHEN OR HOW BIG THEY WILL BE.

EARTH'S FAULTS

Earthquakes occur because of the way big slabs of rock move in Earth's top two layers, the mantle and the crust. These rocks are called tectonic plates, and most earthquakes happen along **faults** on the plates' edges. Sometimes the plates push so hard that they bend rock along a fault. Continued pushing makes the rock break and move suddenly, causing an earthquake.

Earthquakes also happen at soft spots in the middle of plates, but these are not as strong as earthquakes that happen along faults. Earthquakes also happen near volcanoes. Most volcanoes and earthquakes can be found in a path around the Pacific Ocean known as the Ring of Fire.

THE RING OF FIRE

THIS MAP SHOWS THE RING OF FIRE AROUND THE PACIFIC OCEAN. IT ALSO SHOWS SOME OF THE MAJOR TECTONIC PLATES ON EARTH.

LAYERS OF EARTH

FROM THE SURFACE TO THE CENTER OF EARTH, OUR PLANET HAS FOUR MAIN LAYERS: THE CRUST, THE MANTLE, THE OUTER CORE, AND THE INNER CORE. THE CORE IS MOSTLY MADE UP OF IRON. THE INNER CORE IS SOLID, AND THE OUTER CORE IS LIQUID. THE MANTLE IS ROCKY AND MOSTLY SOLID, AND THE CRUST SITS ON TOP.

SPREADING OUT

Earthquakes can begin hundreds of miles beneath Earth's surface. The point where the rock first breaks is called the focus. The epicenter of an earthquake is the spot on Earth's surface right above the focus.

The break spreads along the fault. Sometimes one side of the fault drops below the other side, creating an incline on Earth's surface called a fault scarp. Sometimes one side moves up and over the other side. Other times the two sides slide past each other. The breaking rock lets out energy that travels through Earth in **seismic** waves. Different kinds of seismic waves have different effects on how the ground shakes during an earthquake.

SEISMIC ACTIVITY

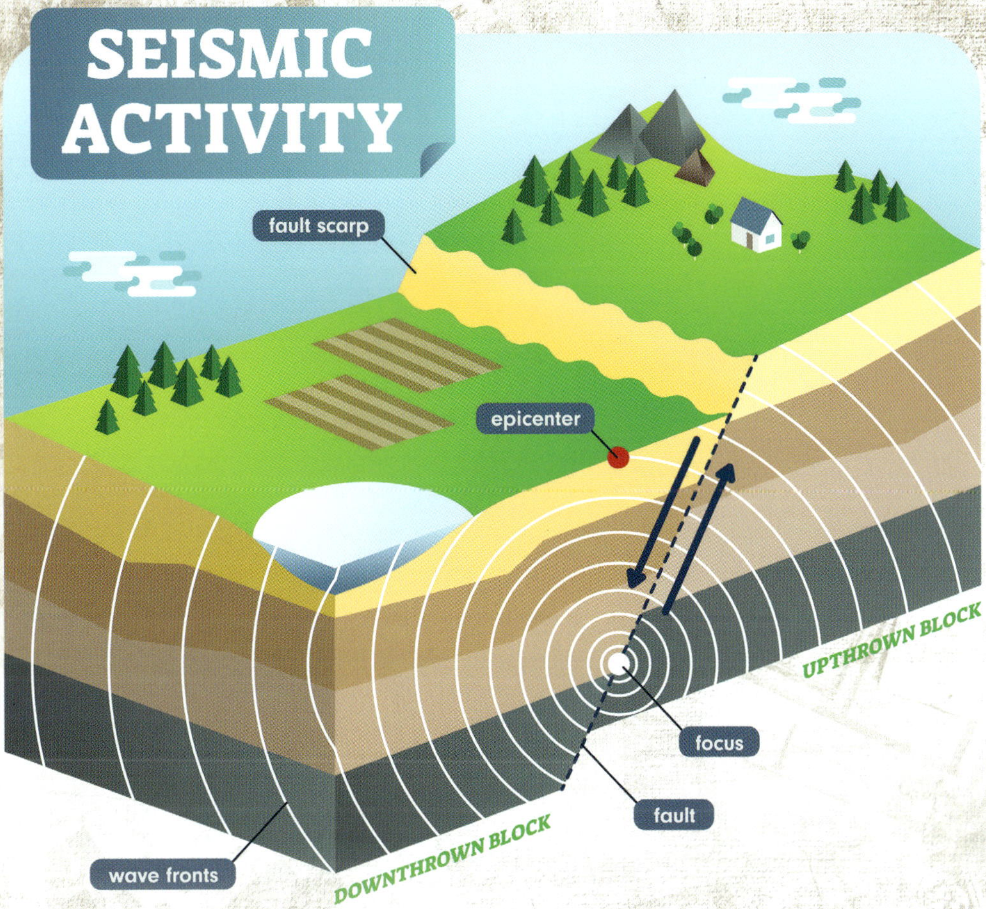

THIS DIAGRAM SHOWS THE FEATURES OF AN EARTHQUAKE. THE SEISMIC WAVES MOVE OUTWARD FROM THE FOCUS INSIDE EARTH.

BODY AND SURFACE WAVES

THERE ARE TWO MAIN TYPES OF SEISMIC WAVES: BODY WAVES AND SURFACE WAVES. BODY WAVES TRAVEL DEEP INSIDE EARTH, WHILE SURFACE WAVES MOVE ALONG THE SURFACE. BOTH WAVES CAUSE THE GROUND TO SHAKE, BUT THEY MOVE AT DIFFERENT SPEEDS. THE GROUND SHAKES IN DIFFERENT DIRECTIONS, SUCH AS UP AND DOWN OR SIDE TO SIDE, DEPENDING ON THE KIND OF WAVE.

EARTHQUAKE DAMAGE

Earthquakes cause different kinds of damage depending on their **magnitude** and location. The breaking rock and seismic waves can tear up structures and roads, shake them, or cause them to fall. Structures can kill people when they fall and break. Earthquakes on the bottom of the ocean can cause tsunamis, which are huge ocean waves. Tsunamis can knock down structures and cause **severe** flooding. People can drown in the tsunamis as well.

There may be no food or clean water after an earthquake. Damaged roads and bridges may make it hard for help to reach people. People may be injured, or they may lose their homes or loved ones.

THE GREAT PERUVIAN EARTHQUAKE OF 1970 CAUSED SEVERE DAMAGE IN PERU. HERE, U.S. FIRST LADY PAT NIXON CLIMBS OVER DEBRIS IN A VILLAGE IN THE ANDES MOUNTAINS.

FURTHER DISASTER

EARTHQUAKES CAN CAUSE DISASTERS OTHER THAN TSUNAMIS. ONE EXAMPLE IS A LANDSLIDE, IN WHICH ROCKS, DIRT, AND OTHER DEBRIS MOVE RAPIDLY DOWN A SLOPE. THE GREAT PERUVIAN EARTHQUAKE CAUSED LANDSLIDES THAT DESTROYED THE TOWN OF YUNGAY AND DAMAGED OTHER TOWNS NEARBY. NEARLY 70,000 PEOPLE DIED. IT WAS THE WORST NATURAL DISASTER IN PERU'S HISTORY.

LEARNING FROM 1906

On April 18, 1906, a strong earthquake struck San Francisco, California. This quake was important because scientists were able to study it and learn more about earthquakes than ever before. At the time, scientists didn't even know about tectonic plates. No one knows for sure how strong this quake was, but some scientists think it had a magnitude around 7.9.

The San Francisco earthquake happened along the San Andreas fault. It started fires that burned for three days. The earthquake and fires destroyed most of the city. About 3,000 people died, and about 250,000 people were left homeless. It was the most famous earthquake in U.S. history and also the deadliest.

HISTORY'S BIGGEST QUAKES

1556

8-MAGNITUDE earthquake in Shaanxi, China. One of the deadliest in history, killing about 830,000 people.

1755

8.5–9.0 MAGNITUDE earthquake in Lisbon, Portugal. This earthquake affected Europe so strongly that people made art and literature about it for centuries.

1950

8.6-MAGNITUDE Assam-Tibet earthquake. This quake caused landslides so big that a natural dam formed, blocking the Subansiri River.

1964

9.2-MAGNITUDE earthquake in Alaska. The strongest earthquake ever recorded in the United States.

1976

7.8-MAGNITUDE earthquake in Tangshan, China. This quake hit in the middle of the night, killing 242,000 to 655,000 people.

⚠ HERE ARE SOME OF THE BIGGEST EARTHQUAKES BEFORE THE 21ST CENTURY. MAGNITUDES FOR EARTHQUAKES EARLIER THAN THE 20TH CENTURY ARE ESTIMATES BASED ON HISTORICAL RECORDS.

THE SAN ANDREAS FAULT

THE SAN ANDREAS FAULT HAS CAUSED EARTHQUAKES FOR CENTURIES, BUT IT WASN'T WELL KNOWN UNTIL THE SAN FRANCISCO EARTHQUAKE. TWO TECTONIC PLATES MEET AT THIS FAULT, WHICH IS ACTUALLY A NETWORK OF MANY OTHER FAULTS. THE SAN ANDREAS FAULT STRETCHES MORE THAN 800 MILES (1,300 KM) AND REACHES AT LEAST 10 MILES (16 KM) DEEP INTO THE GROUND.

THE GREATEST MAGNITUDE

The biggest earthquake that seismologists ever recorded was the Valdivia earthquake on May 22, 1960. It struck southern Chile with a magnitude of 9.5. Around 1,655 people died, and 2 million people lost their homes.

Though this was the biggest earthquake, it was nowhere near the deadliest. This may be for a few reasons. **Foreshocks** warned people a quake was coming. The earthquake also happened in the afternoon, when most people were awake and could get to safety.

The ground sank in some areas, causing floods that changed Chile's coastline. The quake even caused a tsunami that crashed across the Pacific Ocean, affecting Hawaii, Japan, and the Philippines.

PEOPLE EXPLORE A HUGE CRACK THAT OPENED UP IN ONE OF VALDIVIA'S STREETS DURING THE EARTHQUAKE IN 1960.

MEGATHRUST EARTHQUAKES

THE VALDIVIA EARTHQUAKE WAS SO BIG BECAUSE IT HAPPENED ON WHAT'S CALLED A MEGATHRUST FAULT, WHICH CAN BE HUNDREDS OF MILES LONG. ONE HUGE TECTONIC PLATE SLIPPED UNDER ANOTHER, CAUSING THE MEGATHRUST EARTHQUAKE. WE'RE UNLIKELY TO EVER SEE A BIGGER EARTHQUAKE, AS THERE ARE NO KNOWN FAULTS LONG ENOUGH TO CAUSE ONE WITH A MAGNITUDE OF 10 OR HIGHER.

CAUSING A TSUNAMI

A powerful megathrust earthquake happened in the Indian Ocean on December 26, 2004. It occurred off the west coast of the Indonesian island of Sumatra. Scientists call it the Indian Ocean earthquake. Sometimes it is called the Sumatra-Andaman earthquake too.

This earthquake had a magnitude of 9.1. The fault was on the ocean bottom, and the break was about 750 miles (1,200 km) long—nearly the length of the state of California. It caused a huge tsunami, which traveled thousands of miles and hit 17 countries. The tsunami caused most of the deaths and damage. Nearly 2 million people were left homeless, and about 230,000 people died.

THE TSUNAMI DAMAGED BUILDINGS, SHIPS, AND MUCH MORE. THE **DESTRUCTION** SEEN HERE HAPPENED IN THE CITY OF BANDA ACEH IN INDONESIA.

EARTHQUAKE SWARMS

IN JANUARY 2005, A SERIES OF SMALLER EARTHQUAKES HAPPENED NEAR WHERE THE INDIAN OCEAN EARTHQUAKE TOOK PLACE. SUCH A SERIES OF EARTHQUAKES IS CALLED AN EARTHQUAKE SWARM. LIKE THE TSUNAMI, THE SWARM WAS CAUSED BY THE INDIAN OCEAN EARTHQUAKE A MONTH EARLIER. THE SWARM LASTED FOR AT LEAST ONE WEEK AND WAS THE MOST INTENSE EARTHQUAKE SWARM EVER RECORDED.

CHINA IN 2008

China is one of the top countries in Asia for earthquake activity. On May 12, 2008, a 7.9-magnitude earthquake hit China's southwestern province, Sichuan. The quake was so powerful, it could be felt in Beijing, about 1,100 miles (1,800 km) away. People in Beijing said buildings shook for nearly two minutes!

In Sichuan, the damage and loss of life were severe. At least 70,000 people died, 5 million people lost their homes, and more than 21 million buildings were damaged or destroyed in Sichuan and nearby provinces. An estimated 45.5 million people were affected by the quake.

YOU CAN VISIT SOME OF THE RUINS OF BEICHUAN. CHINA PRESERVED THE AREA AS A MEMORIAL TO THE EARTHQUAKE'S VICTIMS.

MAJOR IMPACT

THE EARTHQUAKE HAD A HUGE IMPACT ON MANY DIFFERENT PARTS OF CHINA. IT CAUSED AROUND 200,000 LANDSLIDES, WHICH DESTROYED CITIES AND BLOCKED RIVERS. ROADS, DAMS, AND DAMAGED TAP WATER PIPELINES. ONE DESTROYED CITY, BEICHUAN, WAS NOT REBUILT. IT BECAME A GHOST TOWN, WITH PLANT LIFE SLOWLY GROWING OVER THE RUINED BUILDINGS.

EARTHQUAKES IN HAITI

Haiti has faced powerful earthquakes for centuries. This is because the country sits between two fault zones, or groups of faults. There's one zone in the north and one in the south.

On January 12, 2010, the deadliest earthquake in Haiti's history struck near the country's capital, Port-au-Prince. With a magnitude of 7.0, the earthquake caused a lot of damage. More than 1 million people lost their homes. Haiti's government reported that 220,000 to 316,000 people died, making this the deadliest earthquake in modern history. Because the damage was so severe, it was hard to know exactly how many people were affected.

PORT-AU-PRINCE WAS BADLY DAMAGED AFTER THE 2010 EARTHQUAKE. THE CITY HAS BEEN REBUILT SEVERAL TIMES, INCLUDING AFTER EARTHQUAKES IN 1751 AND 1770.

THE 2021 QUAKE

ON AUGUST 14, 2021, AN EARTHQUAKE WITH AN EVEN LARGER MAGNITUDE HIT HAITI. THIS 7.2-MAGNITUDE EARTHQUAKE DAMAGED ABOUT 13,000 HOMES, AND 2,200 PEOPLE DIED. EVEN THOUGH IT WAS A BIGGER EARTHQUAKE, THERE WAS NOT AS MUCH DAMAGE AS THERE WAS IN 2010. THIS MAY BE BECAUSE THE EARTHQUAKE HAPPENED FARTHER AWAY FROM WHERE PEOPLE LIVED.

JAPAN IN 2011

Japan is located within the Ring of Fire, and it has experienced many earthquakes throughout its history. Scientists have set up many seismographs and other tools to track seismic activity there. On March 11, 2011, they detected tremors and warned the public an earthquake was about to begin.

The 9.0-magnitude earthquake happened off the eastern coast of Japan's Tōhoku region. It caused a tsunami that destroyed homes and hit a **nuclear** power plant. This caused a nuclear **meltdown**, which leaked harmful materials into the environment. Thousands of people had to leave the area. With at least $220 billion of damage, this natural disaster cost more than any other in world history.

HERE, A SEARCH TEAM AND TRAINED RESCUE DOG LOOK FOR MISSING PEOPLE ON MARCH 15, 2011, FOUR DAYS AFTER THE EARTHQUAKE IN JAPAN.

EARLY WARNING SYSTEM

JAPAN'S EARLY WARNING SYSTEM ALLOWS SCIENTISTS TO ALERT PEOPLE AN EARTHQUAKE IS ABOUT TO BEGIN. THIS IS DIFFERENT FROM PREDICTING EARTHQUAKES LONG BEFORE THEY HAPPEN, BUT THE SYSTEM CAN GIVE PEOPLE UP TO A MINUTE OF WARNING TO GET TO SAFETY. THIS CAN BE LIFESAVING. ALTHOUGH MORE THAN 15,500 PEOPLE DIED IN THE 2011 QUAKE, MANY OTHERS WERE SAVED THANKS TO THE WARNING.

THE WORST OF 2023

Türkiye (formerly Turkey) is another country that has seen many powerful earthquakes. On February 6, 2023, a 7.8-magnitude earthquake struck near the border between Türkiye and Syria. It was the strongest to hit Türkiye in nearly a century, and it was one of the deadliest of the 21st century so far. At least 50,000 people died, and more than 100,000 others were injured.

The quake occurred early in the morning, but **aftershocks** continued for a few weeks. The aftershocks and cold February weather made the affected areas more dangerous. Thousands of buildings were damaged or destroyed across southern Türkiye and northwestern Syria, and around 1.5 million people lost their homes.

THIS PHOTO SHOWS SOME OF THE DESTRUCTION IN HATAY, TÜRKIYE. MANY BUILDINGS FELL STRAIGHT DOWN, IN WHAT IS CALLED A "PANCAKE" COLLAPSE.

STRIKE-SLIP FAULT

THE EARTHQUAKE IN TÜRKIYE HAPPENED ON A STRIKE-SLIP FAULT. THIS MEANS THAT TWO TECTONIC PLATES SLID PAST EACH OTHER SIDEWAYS, MAKING THE EARTH SHAKE FROM SIDE TO SIDE. THIS KIND OF MOTION CAN CAUSE MORE DAMAGE TO BUILDINGS THAN OTHER KINDS OF EARTHQUAKES, WHICH IS PARTLY WHY SO MANY BUILDINGS COLLAPSED.

EARTHQUAKE SAFETY

There are things you can do to stay safe when an earthquake happens. If you're inside, get under a strong table or desk. This will keep you safe from anything that might fall during the quake. Stay there until the shaking stops. If you're outside, get as far away as you can from utility poles, buildings, and trees.

Seismologists know where earthquakes may happen, but not when. If you live in an area where strong earthquakes happen, talk to your parent or guardian about how you can be prepared. Having a disaster plan can help keep you and your family safe in the event of an earthquake.

PRACTICE "DROP, COVER, AND HOLD ON" WITH YOUR FAMILY. DROP TO THE FLOOR, TAKE COVER UNDER A STURDY TABLE, AND HOLD ONTO IT UNTIL THE EARTHQUAKE STOPS.

BE PREPARED

THERE ARE PLENTY OF STEPS YOU CAN TAKE TO PREPARE FOR AN EARTHQUAKE. MAKE SURE HEAVY OBJECTS ARE LOW TO THE GROUND INSTEAD OF ON HIGH SHELVES. CREATE AN **EMERGENCY** BAG WITH A FIRST-AID KIT, A FLASHLIGHT, AND OTHER SUPPLIES. TALK TO YOUR FAMILY ABOUT HOW TO CONTACT EACH OTHER OR WHERE TO MEET IF YOU GET SEPARATED DURING AN EARTHQUAKE.

WORKING TOGETHER

When a major earthquake happens, governments and disaster relief teams work together to help people who are affected. Officials also do what they can to make things safer the next time a quake hits. After the 2011 Tōhoku earthquake proved Japan's early warning system was effective, the government of Chile worked with Japan to create an early warning system of its own.

In the United States, seismic codes help ensure current and future buildings are made to withstand earthquakes. These codes are part of a set of building rules that engineers use. The codes are updated when new findings show how to make buildings stronger and more flexible, creating a safer future for everyone.

ENGINEERS AROUND THE WORLD CONSIDER EARTHQUAKE RISK WHEN BUILDING STRUCTURES. THE ENGINEER SEEN HERE PREPARES TO TEST A BRIDGE'S STRENGTH FOR THE EUROPEAN CENTER FOR EARTHQUAKE ENGINEERING IN ITALY.

HELPING HANDS

DISASTER RELIEF TEAMS ARE USUALLY MADE UP OF VOLUNTEERS, PEOPLE WHO DONATE THEIR TIME TO HELP. ORGANIZATIONS SUCH AS THE RED CROSS OFFER TRAINING SO VOLUNTEERS KNOW WHAT TO DO WHEN A DISASTER HAPPENS IN THEIR AREA. IN THE UNITED STATES, THE FEDERAL EMERGENCY MANAGEMENT AGENCY (FEMA) OFFERS TRAINING FOR VOLUNTEERS THROUGH THE COMMUNITY EMERGENCY RESPONSE TEAM PROGRAM.

GLOSSARY

aftershock: Minor earthquake that follows the major shock of an earthquake.

damage: Harm. Also, to cause harm.

destruction: The state of being destroyed or ruined.

disaster: Event that causes much suffering or loss.

emergency: An unexpected situation that needs quick action.

fault: Crack along the edge of a tectonic plate.

foreshock: Tremor that happens before the main shock of an earthquake.

magnitude: A measure of the power of an earthquake.

meltdown: An accident that causes a nuclear reactor core to overheat.

nuclear: Having to do with the power created by splitting atoms, the smallest bits of matter.

predict: To guess what will happen in the future based on facts or knowledge.

seismic: Having to do with an earthquake.

severe: Very bad or serious.

FOR MORE INFORMATION

BOOKS

Palffy, Georgina. *Knowledge Encyclopedia Planet Earth!* New York, NY: DK Publishing, 2022.

Romero, Libby. *All About Earthquakes.* New York, NY: Children's Press, 2022.

Tarshis, Lauren. *I Survived the Great Alaska Earthquake, 1964.* New York, NY: Scholastic, 2023.

WEBSITES

National Geographic Kids
kids.nationalgeographic.com
Learn more about the science of how earthquakes happen and affect our world.

Ready Kids
ready.gov/kids/disaster-facts/earthquakes
Learn how to stay safe before, during, and after an earthquake.

USGS: Earthquakes for Kids
earthquake.usgs.gov/learn/kids/
Learn more about the history and science of earthquakes through photos, videos, maps, and more.

INDEX